W9-CFB-889

WE CAN READ about NATURE!™

GRASSY LANDS

by CATHERINE NICHOLS

BENCHMARK BOOKS

MARSHALL CAVENDISH
NEW YORK

*With thanks to
Peggy C. Hansen, Teacher,
Noxon Road Elementary School,
New York, for providing the activities in
the Fun with Phonics section and
to Beth Walker Gambro, Reading Consultant.*

Benchmark Books
Marshall Cavendish
99 White Plains Road
Tarrytown, New York 10591-9001

Photo research by Candlepants Incorporated

Cover Photo: Corbis / Wolfgang Kaehler

The photographs in this book are used by permission and through the courtesy of;
Corbis : Fotographia, Inc., 5; Robert Holmes, 6-7; Annie Griffiths Belt, 8-9;
Tom Bean, 10-11, 26-27; Perry Conway, 12, 17; Darrell Gulin, 13; Kevin Schafer, 15;
Wolfgang Kaehler, 18-19; Ralph A. Clevenger, 20; Robert Gill / Papilio, 21;
Kennan Ward, 22, 23; Craig Aurness, 24; Philip Gould, 25; Julie Habel, 29.

Library of Congress Cataloging-in-Publication Data

Nichols, Catherine.
Grassy lands / by Catherine Nichols.
p. cm. — (We can read about nature!)
Includes index.
Summary: Discusses the different kinds of grasslands, such as prairies, steppes, and savannas, and introduces some of the animals that live there.
ISBN 0-7614-1435-5
1. Grasslands—Juvenile literature. [1. Grasslands. 2. Grassland
ecology. 3. Ecology.] I. Title. II. Series.
QH87.7 .N53 2002
577.4—dc21
2002005012

Printed in Hong Kong

1 3 5 6 4 2

Look for us inside this book.

cattle
cheetah
elephant
lion
owl
prairie dog
red fox
zebra

Grass is all around us.

It grows in parks and lawns.

Grass also grows in the wild.

A large area of grassy land is called
a grassland.

A prairie is a kind of grassland.

Wildflowers bloom in prairies.

Only a few trees grow there.

A steppe is another kind of grassland.

Steppes get very little rain.

That is why the grass is short.

Many animals live in prairies
and steppes.

Owl

Red fox

Prairie dogs live in prairies.

They are not really dogs.

Prairie dog

Prairie dogs are a kind of squirrel.

They live in burrows.

Burrows are holes in the ground.

Savannas are large, flat grasslands.

Trees and bushes grow there
in clumps.

18

Savannas are hot all year.

These animals live in savannas.

Zebras

They eat grass and leaves.

Elephant

Other animals eat the grass-eaters.

But they do not eat the elephant.

Cheetah

That grass-eater is just too big!

Lion

Many grasslands have been lost.

Some have become wheat fields and cornfields.

Cattle and sheep graze on other grasslands.

People are trying to protect
the grasslands.

They plant grass seeds in prairies
and savannas.

Grasslands are beginning to grow again.

These grassy lands can teach us about nature.

It is important to take good care of them.

Fun with Phonics

(Answers on page 32)

1. WHICH WORD DOES NOT BELONG?

Most of the words in each line mean almost the same thing. One word does not belong with the others. Which word does not fit?

> A. prairie, stop, savanna, steppe
> B. burrow, tunnel, hall, hole
> C. grow, blossom, draw, bloom
> D. area, space, spoon, land

2. IT'S A SECRET!

Write the answers to each question on a separate piece of paper. The answers are all words in the story. Circle the first letter of each answer. Together they spell a secret word!

> A. On the prairie, only a few trees are able to do this.
> B. Steppes get very little of this.
> C. Cows, lions, and sheep are different kinds of these.
> D. This place has short grass and few trees.
> E. This place is flat and hot.
> F. These grow on trees.
> G. This word means "once more."
> H. We Can Read about this.
> I. This word goes with prairie to name a kind of squirrel.

30

3. CHALLENGE: RHYME TIME

Read the clues to find two rhyming words. (Example: For the clue "The kitten was not standing" the answer is "cat sat.")

   A. You only have to wipe the table with a soft cloth.
   B. You should chew your hamburger.
   C. Lambs take naps.
   D. Cows tell on each other.

Fun Facts

- Almost one-third of Earth's land is grassland.
- Grasslands are found on every continent except Antarctica.
- Prairies and steppes are temperate grasslands. That means they have hot summers and cold winters.
- The Great Plains of North America is a prairie. It stretches from Canada to Texas.
- Savannas are tropical grasslands. They stay hot all year.
- Cheetahs live in savannas. They are the fastest animals on land. Cheetahs can reach speeds of over 60 miles an hour.
- Elephants also live in savannas. They are Earth's largest living land animals.

Glossary/Index

About the Author

Catherine Nichols has written nonfiction for young readers for fifteen years. She works as an editor for a small publishing company. She has also taught high-school English. Ms. Nichols lives in Jersey City, New Jersey, with her husband, daughter, cat, and dog.

Answers to pages 30–31:

1. Which Word Does Not Belong?
A. stop B. hall C. draw D. spoon
2. It's a Secret!
A. grow B. rain C. animals D. steppe E. savanna F. leaves G. again
H. nature I. dog **Secret word:** grassland
3. Challenge: Rhyme Time
A. just dust B. eat meat C. sheep sleep D. cattle tattle

32